OMNIPOTENT SPACE

THE BASE & THE CAUSE FOR THE UNIVERSE

I0471565

SUBHASH MARAMRAJU

XpressPublishing
An imprint of Notion Press

XpressPublishing
An imprint of Notion Press

Old No. 38, New No. 6
McNichols Road, Chetpet
Chennai - 600 031

First Published by Notion Press 2019
Copyright © Subhash Maramraju 2019
All Rights Reserved.

ISBN 978-1-64650-216-5

Contents

Preface   v

PART I

1. My Approach Towards The Beginning Of The Universe   3
   And The Goal Of Human Life

2. Electronic Era Of Invisible And Subtle Material And   7
   Material Particles

3. Space Is Not Empty   9

4. Virtual Particles   10

5. Quantum Theories   11

6. Big Bang Theory   13

7. Planets   14

8. Milky Way   16

9. Universal Things Composed With 5 Basic Elements   17

10. Atoms & Cells   18

11. Living And Non-living Things   19

12. Subtlety Of Material Particles In The Universe   21

13. Omnipotent Space   24

PART II

14. At The Outset, I Salute To All The Great Legends Born 27
    On The Planet Earth.

15. The Human Body   29

16. The Realization Of The Truth   34

17. How To Approach Towards Realization Of The Truth?   36

18. How To Live After Realizing The Above Truth   38

Gist Of The Book   41

Formation Of The Universe   45

Preface

Subhash Maramraju - MA, BL

At the outset, I wish to introduce myself in brief: I was born on 6th July 1948 at my native place Mahabubabad, now the district headquarters in the State of Telangana (India), where I studied up to XII class equivalent to Intermediate and also learnt Pitman's English Shorthand and typing. I had worked in a Central Government department at Hyderabad from 1968 to 1991 and retired voluntarily as Assistant. While working there, I had continued my studies and completed BA, MA(Public Admn) & LL.B externally from Osmania University. Later, worked as a stenographer in private institutions and as an Advocate in civil courts at Hyderabad for about 10 years altogether. As I was more interested in the fine arts – Literature, Music and Painting. I left all the activities and concentrated on oil paintings. I made visualized oil paintings on historical monuments in the districts of Warangal and Mahabubnagar between 2001-2004 for which I was felicitated and awarded. I had presented the portraits to the President & Vice-President of India, Governor& C.M of Telangana, and to many other celebrities from the academic, film and political fields. Made creative oil paintings and story boards for many people. Recently, I worked temporarily in the administrative department of Osmania University College for Women, Koti, Hyderabad from March 2013 to January 2019. I had conducted exhibitions of my paintings in Dallas (US) & Hyderabad (TS-India) during 2009 & 2017 respectively. Still, I have been working on the paintings now and then.

As a matter of fact, since for the last about four decades, I have been curious to know as to how all the Universe has been created in the infinite Space. Whenever I found leisure time, I had been pondering over the same, attending to the discourses of eminent philosophers, reading their books and also the books of eminent scientists of the world on the creation of the universe. My salutations to all those legends, whose scientific theories and philosophical discourses have been downloaded into my brain, enabling me to make further research into the generation of the material particles from gross to subtle levels, etc, before the formation of the universe. After about an year of persistent study, I could write this small book titled 'Omnipotent Space (The Base & the Cause for the Creation)', to enlighten the people about the true concept of the Space and bring Oneness of all the things (Living and Non-living) in the Universe.

Subhash Maramraju, MA BL
9248691134
(maramsubhash@gmail.com)

PART I

Omnipotent Space

CHAPTER ONE

My Approach Towards the Beginning of the Universe and the Goal of Human Life

I have started my approach from the beginning of the generation of subtle particles of things from the space, much before the formation of the stars, planets and the five main elements – Earth, Water, Fire, Air & Ether in the cosmos.

I needed deeper study and research to find out a proper cause for the creation of all the living and non-living things in the universe. As I studied from various material and theories, I found that many Astronomers/scientists use empty space to study the universe in respect of materialistic things, which are being accessible to human mind & body, ignoring the Space as being empty, void and nothingness.

But, some scientists say empty space is containing dark energy, which is causing the expansion of the universe to accelerate and that space is fully occupied by Ether, black energy, black matter, black holes, void & nothingness. Occupation of all these things requires the space to be occupied. Thus, everything in the universe is existing in the Space, which every human being knows as only empty place accommodating all the material things.

SPACE is all pervading, infinite and eternal without any form or color or smell, but it creates various magnetic, gravitational forces to keep various galaxies intact. It exists IN and OUT of all the end particles of the cells/atoms of the Universal things we see. From the following paras, we realize that it is not only the base but also the generator, organiser and destroyer of all the materialistic things of nature.

Recently, a thought was flashed in my mind and I wanted to study on the subject further and establish that *'everything in the Universe must have been generated from the infinite Space itself since there is no other source for the humans to think about. Because, Space is infinite, eternal and it was only the existing energy all over, in which we see the whole universe as is being witnessed now. It is also obvious and understandable to every human being that there is no further base for the infinite Space.*

Another aspect I thought was, that *'everything must have been generated or born in an invisible and subtle form and later grew up to a level of the visible object, as is evident from our birth itself.'* This phenomenon is possible only in the invisible, all-pervading, formless and eternal base, which must not have any further base for it, but at the same time, it must also have the abundant energies required to generate the subtle particles of various elements in the universe, as we see it now.

Being an artist since my childhood, I visualized a picture of the state of Cosmos as it would have been looked much before the occurring of the Big Bang and the formation of the Stars, Planets, etc in the sky as detailed by the scientists. The picture I saw in my mind was a vast blank area of darkness covered with billions of tiny white dots when deeply observed; and there were all serenity and silence prevailing all over, creating a blissful feeling in my mind. It was the Space with energy particles. Then, I wanted to go step by step by logic, reasoning and cause & effect for the generation of all the material things from their state of the subtle level to the gross level.

As you are aware, the eminent scientists have discovered and invented many things to facilitate and develop the human society to lead a healthy, comfortable and happy life. People of this generation have also become so fast and completing their day to day works through wonderful electronic gadgets & internet in a matter of seconds. The moment I saw a Smartphone, which was invented in the recent past, I became very much interested to make a research into the generation & flow of such subtlest particles in the Space all over the Universe. Though I am not a professional scientist, I cherished to know about the formation of the Universe. In this connection, I have studied various scientific theories about the Space and generation of various subtle particles, their growth, and development of their energy levels, expansion of the universe as we see it today, etc.

Later, being satisfied with my approach about the Space, I am writing my thoughts into a small book to enlighten the human society that Space is not only the Base but also the Cause for the creation of all the things in the Universe. This is very useful to all human beings to know the truth, get wisdom and live their full life happily and peacefully.

Space is known to every human being in the world as an empty space, void, vacuum, and nothingness, which is not true. After going through the foregoing points based on cause and effect, logic and reasoning, one must be convinced that it is an infinite, powerful Space, being the Base and the Cause for the creation of all the things in the Universe. It is easy to understand, except for the gist of some scientific theories, which I added only to support my thesis. It is enough when you become engrossed in the substance while reading this book, you would come to a conclusion that you cannot imagine anything in the universe without Space, which occupies more than 99.9999999% of an atom, while everything is formed with the composition of millions of atoms.

Realization of the above truth, would in no way obstruct and deviate you from doing your duties and responsibilities towards your family, and in fulfilling to reach your desired goals of mundane life. Over and above, you will get wisdom, strength, energy, and enthusiasm after you understood the truth that the Space existing within your whole body and concurrently in all the things of the universe, is One and the same.

CHAPTER TWO

ELECTRONIC ERA OF INVISIBLE AND SUBTLE MATERIAL AND MATERIAL PARTICLES

i. The First question that arose in my mind after seeing the Smartphone, was *'what are those invisible tiniest particles, which carry the videos and audios from one place to another, of long distances all over the world instantly within seconds?'*

ii. The Second question was, *' when an invisible Atom is said to be a unit for all material things, what about its sub-particles inside viz Electrons, Protons, Neutrons, quarks, leptons ,gluons, etc?'*. Then logically I thought, that every thing could be reduced to a size till it is totally immersed in the Space.

iii. Third question was, *'When a thing finally immerses into the Space, why not anything could generate from out of the space, whatever invisible, tiniest particle or energy it be?'*.

iv. Fourth question was that, '*what was the first invisible material element that was born out of the Space to create all the material in the Universe?*'

Thus, I started making research into the beginning of the universal material particles as per the gist of the Scientific information in the next few sections

CHAPTER THREE

SPACE IS NOT EMPTY

In the recent past some eminent scientists of the world, after continuous research, discovered that free Space or Vacuum is not empty, but seething with energy & gravity. It will also be understood in the ongoing paras, that the vast Space has also subtle energies of intelligence and the consciousness. It could also yield flashes of light with a pair of particles popping in and popping out of the infinite Space all over, intermittently. Since these particles do not have permanent existence, scientists called them as "Virtual particles" or fluctuations of vacuum energy.

CHAPTER FOUR

VIRTUAL PARTICLES

Space is not at all empty because *Nothing contains something*, exasperated with energy and a pair of virtual particles (particle & antiparticle) that flit into and out of existence. They are also called vacuum energy and its fluctuations. It was also discovered that those virtual particles could turn to become real particles and come into permanent existence. This effect is known as the Schwinger effect, which was supported by other cosmologists that, under the strong gravitational field, some virtual particles could be made to become real photon particles. These photons create light waves. Now, question arises in our mind as to how the photon particles created such a big Universe. The infinite Space is also filled with the energies of main elements after their annihilation from time to time. Thus, in the infinite Space not only the virtual particles but the particles of other elements must have also been generated all over and accumulated at various points of the space during a long period of time before the Big Bang had occurred. Further, the following Quantum theories explain about the matter and energy created and expanded in the space.

CHAPTER FIVE

QUANTUM THEORIES

Quantum theory is the theoretical basis of modern physics that explains the nature and behavior of matter and energy on the atomic and subatomic level, as detailed in its theories as follows:

(i) Quantum physics: It explains that there are limits to how precisely one can know the properties of the most basic units of matter—for instance, one can never absolutely know a particle's position and momentum at the same time. One bizarre consequence of this uncertainty is that a vacuum is never completely empty, but instead buzzing with so-called "virtual particles and photons" that constantly wink into and out of existence leaving behind some energy. Still, before they vanish, they can have very real effects on their surroundings. Photons can pop in and out of a vacuum. They have mysterious force. For example, when two mirrors are placed facing each other in a vacuum, more virtual photons can exist around the outside of the mirrors than between them, generating a seemingly mysterious force that pushes the mirrors together.

(ii) Quantum mechanics: It is the science of the very small particles. It explains the behavior of matter and its interactions with energy on the scale of atoms and subatomic particles. Light behaves in some aspects like particles and in other aspects like waves. Matter—the "stuff" of the

universe consisting of particles such as electrons and atoms exhibits of particles wavelike behavior too. A single photon is a quantum, or smallest observable amount of the electromagnetic field because a partial photon has never been observed.

(iii) Quantum Field Theory: *According to the experimental observations, the negative charged particle produces positive virtual photon and positive charged particle produces negative virtual photon. A negative and a positive virtual photon combine with each other in the vicinity of a charged particle cause the charged particle to accelerate in space.*

Thus, the above Quantum theories also explain how the virtual particles behave and become the real photons, including their structure, force and energy particles speeding up in the space.

The scientists discovered that these virtual particles which became real photons, their electrons and protons (negative and positive charge) interactions, and their chemical reactions with the lightest elements of Hydrogen and Helium, Lithium and Beryllium, which were similarly generated in the Space, created heavy energy and gravity, causing a First Explosion in the Cosmos. This was named as 'Big Bang'.

CHAPTER SIX

BIG BANG THEORY

About 13.8 billion years ago, as stated above, the Big Bang had occurred in the cosmos and it was continued for more than a half billion years. After this, 1.6 million years later, gravity began, clumps of gas collapsed enough to form the Stars and galaxies from clouds of gas. In the *first* moments, after the *Big Bang*, the universe *was* extremely hot. It has been expanding ever since, to its current size of something like 100 billion galaxies. It had not occurred at a particular point in the space. Such explosions had been occurring throughout the space.

The first Stars formed in the cosmos, had created the new elements in their cores by squeezing elements together, in a process called Nuclear Fusion. Initially, stars fused hydrogen atoms into helium. Helium atoms then fused to create Beryllium and so on, until the fusions in the stars' cores had created every element up to the element of iron. Ever since the **Big Bang**, the Universe has been drifting and expanding. The birth and death of the Stars left an aftermath of galaxies, with *planets*, Comets, Asteroids, including living organisms. Roughly 3,80,000 years after the Big Bang, the matter cooled enough for atoms to form.

CHAPTER SEVEN

PLANETS

First star was called **SUN,** which was made up mostly of hydrogen, helium and other elements, and the other stars that came after contained heavier elements such as carbon and iron. From the hot and burning Sun some explosions had occurred and various objects were exploded and scattered across the space around the Sun. These objects started revolving around the Sun at a particular distance, on account of Gravitational force of the Sun. These objects are named Planets. The gravity of the Sun keeps the planets in their respective orbits. They revolve in their orbits around the Sun, because there is no other force which can stop them.

OUR **EARTH** is one of the observable 9 Planets as far as we have been aware of, but in the space there may be millions of such planets. The research is still going on to find out more planets in our galaxy. Earth was formed around 4.54 billion years ago, approximately one third the age of the universe and since then it has been rotating on its own axis, causing day and night and also revolving around the Sun, causing the seasons. The other objects created from explosions in the Sun are Comets and asteroids, which contained ice and colliding with Earth, they delivered oceans of water, as opined by scientists. During the time immemorial, the remaining elements Fire and Air developed on Earth gradually, out of chemical combinations of various molecules of Hydrogen,

Helium, Carbon, Nitrogen, Oxyzen, iron, calcium, potassium, phosperous, etc. One smaller planet revolving around our Earth was named as **MOON**.

CHAPTER EIGHT

MILKY WAY

All the objects we see from Earth viz Sky, Sun, Moon, Stars, Planets, comets and asteroids combined, are called a Galaxy, named as **Milkyway**, which is our Galaxy. The scientists are still probing into this galaxy for discovering new planets, etc. There are billions of such unnamed galaxies in the infinite Space, which are unknown to the scientists till now.

CHAPTER NINE

UNIVERSAL THINGS COMPOSED WITH 5 BASIC ELEMENTS

All of us are aware of the fact that, the whole universe is broadly consisting of 5 Elements, named from gross to subtle levels as Earth, Water, Fire, Air and Ether (Aakash). Ether is a medium generated in the Space to propagate and carry light and sound waves in the Space. The most commonly observed states of all the material things categorized are in the state of solid, liquid, gas and plasma, which share many attributes with the classical elements of earth, water, air, fire and Ether. The state of Plasma is a hot ionized gas consisting of approximately equal numbers of positively charged ions and negatively charged electrons. The characteristics of plasmas are significantly different from those of ordinary neutral gases so that plasmas are considered a distinct "fourth state of matter". Plasma is a neon sign. Just like a fluorescent lights, neon signs are glass tubes filled with gas. Stars are a good example of how the temperature of plasmas can be very different. The electrons, having a wave function in a chunk of atom, occupy everywhere in the space along with "Ether", the subtlest fifth element, which is the wave based light to propagate through empty space as a medium in the space.

CHAPTER TEN

ATOMS & CELLS

An Atom is considered as a unit of various things formed out of the above said 5 elements. But the Atom includes many sub-particles in it as detailed below:

Of all the material objects of both living and non-living things, the tiniest invisible unit was named as 'Atom". Composition of 2 or more atoms with different molecules, eg. 2 Hydrogen atoms and 1 Oxygen atom forms a cell of water (H_2O). Atoms are composed of sub-atomic particles called Electrons, Protons and Neutrons, quarks, leptons , gluons, etc. These are made of even tiniest particles called, gluons, quarks. protons and neutrons are in the center of the Atom, making up the nucleus. An **ion** is a **charged** atom or molecule. When an atom is attracted to another atom because it has an unequal number of **electrons and protons**, the atom is called an **ION**. Electrons are negatively charged, Protons are positively charged and Neutrons are not charged. If the atom has equal number of Electrons and Protons the charges are cancelled and the net charge the Atom gets is zero. Tiny particles called quarks and *gluons* are the building blocks for larger particles such as protons and neutrons, which in turn form atoms.

The above points reveal that there are more tiniest particles than the Atom and its sub-particles. It is also obvious that the Space is the Ultimate Base for all the last subtlest particles of all the elements to generate from, to stay in, to grow and immerse in it when they become weak and die.

CHAPTER ELEVEN

LIVING AND NON-LIVING THINGS

All the things in the universe have been categorized into TWO – (i) Living things (Male & Female) & (ii) Non-living things. Non-livings, which are static and lifeless, are formed by the Atoms viz earthen material elements, mountains, rivers, gold., stones, iron, wood, including dead bodies of all living things, etc.

Living things with life are formed by Cells. The Cells are formed through combination of two or more atoms of different Molecules (Oxygen, Hydrogen, Nitrogen, Helium, Phosphorous, Calcium, etc, etc). According to the various combinations of the atoms and molecules, billions of various living things (both invisible and visible) are born. Viz Bacteria, Insects, aquatic creatures, birds, animals, trees and human beings, both male and female, which born, grow, move, take respiration and have reproductive system, have brain and consciousness. All the Cells are created with multiple combinations of all the 5 Elements.

Of all the living things, human beings have a super Brain in the skull of the head of the body which has invisible and subtlest particles called neurons and through these emanate the mysterious mind of thoughts and intelligence with conscience and consciousness of all the sensitive organs of the whole body.

Thus, more than a billion (100 crore) varieties of Living and Non-living things in the Universe, now we see, are the result of the Quantum theory and Reproductive system (the production of offspring by a sexual or asexual process among male and female living things & various combinations and chemical reactions of the atoms of various non-living things) and chemical actions and reactions across the above said elements. For example the ten numbers 1,2,3,4,5,6,7,8,9,0 could be arranged into billions of different cell phone numbers for all the people in the world, by way of their variations in placements, jumbling arrangement, duplication, etc. Similarly the various particles of different properties turned into billions of living and non-living things by physical, chemical, biological variations and through reproduction of living organisms.

CHAPTER TWELVE

SUBTLETY OF MATERIAL PARTICLES IN THE UNIVERSE

1. All the things formed through the 5 broad elements of the universe originated from subtlest particles and these are in Solid, Fluid, Gas & Plasma forms. As stated above, an invisible Atom is the unit of everything. But what about the Nuclei, Electrons, Protons and Neutrons existing in that Atom. So there are many unimaginable tiniest particles exist, but could not have been discovered so far. Hence the fact is that all the subtlest particles of the universal things must have been generated like Virtual particles from the infinite free Space and later becoming the real particles.. The elementsfrom gross level to subtle level immerse in one another viz - Earth dissolves in Water, Water in Fire, Fire in Air, Air in Ether. Of these 5 elements, Ether, being the subtlest element, is spreading all over the Space as a space-filling substance. As per scientists Ether, composed of some energy and waves, act as transmission medium for the propagation of electromagnetic or gravitational forces, thus carrying sound and light waves across the space all over. Subtler than Ether there are 3 special elements in the

human brain, which are the thoughts of Mind, Intelligence and false ego. Including these 3, the total elements in the Universe could be said to be Eight (8).

2. When you are able to concentrate on a particular subtle thing, the other thoughts of mind vanish and your intelligence stands on that particular thing, when you get a subconscious feeling "I am"(the false ego),which is the base and subtler than the intelligence. If you meditate upon the vast Space itself, the intelligence and false ego vanish (merge into the Space) since there is no further base for the infinite Space. Thus you will become permanently detached with all the material life at that moment.

3. As science reveals that every Atom is occupied by 99.9999999% of the space, a doubt arises in us: as to why all the things look and feel solid when touched, though the Space occupies 99.9999999% of all the things in the universe.

i.   As per the science, this is all because of electrons revolving around the nucleus in an atom at some distance. While moving faster they take energy and when they move in a slower pattern lose energy and throws in the form of light and when this energy particles fall on another electron, it absorbs the same and move up to a higher and faster speed in a dance pattern. Thus electrons in all the atoms try to grab the energy from light, which is a cause for the thing looking solid.

ii. Another clarification is that, if you touch the table, then the electrons from atoms in your fingers become close to the electrons in the table's atoms. As the electrons in one atom get close enough to the nucleus of the other, the patterns of their dances change. This is because, an electron in a low energy level around one nucleus can't do the same around the other – that slot's already taken by one of its own electrons. The newcomer must step into an unoccupied, more energetic role. That energy has to be supplied, not by light this time but by the force from your probing finger. So pushing just two atoms close to each other takes energy, as all their electrons need to go into unoccupied high-energy states. Trying to push all the table-atoms and finger-atoms together demands an awful lot of energy – more than your muscles can supply. You feel that as resistance to your finger, which is why and how the table feels solid to your touch. (There are many theories from Chemistry, Quantum theories, etc to prove this aspect).

iii. The above facts clearly reveal that the stars, planets, asteroids, comets and all the living and non-living things generated out of the vital elements of earth, water, fire, air and ether on the planet Earth, have been generated from the Space and existing in the Space and ultimately destroy in the Space when their respective subtle particles become weak and loose their energies into the Space. Thus the Space generates things temporarily for a certain period, destroy them and take back their energies and keep in Itself. Thus, we can see only the Space in all the universal things. When you imagine yourself to be far away from the stars and planets, you observe vast dark space all over except a small portion with billions of tiny light spots – stars, planets, etc. You cannot see our Earth at all.

CHAPTER THIRTEEN

OMNIPOTENT SPACE

The Space is the ultimate infinite state of subtleness, without any beginning and ending, and has no further base for it, and hence it is able to store all the subtle and invisible energies of the universal things in it, occupies itself in all those energies and generate them as virtual particles, creates the explosions like Big Bang and form the universe as we see it today. We do not know since what period of time such activities of the Omnipotent Space had been going on and how long this process would be continued in future. Scientists opine that already there must be billions of galaxies in the infinite Space, which we are unable to find them from our planet 'Earth'.

All the energies of the universal elements combined, are far lesser than the energy that the Space has. Thus, without Space we cannot imagine anything in the Universe.

For easy understanding, I name the space occupied in all the things of the universe as **'Inner Space'** and the space all over the cosmos as **'Infinite Space'**. Both are one and the same i.e. the '**Omnipotent Space', *the Base and the Cause for the Universe.***

PART II

HUMAN BEING – WAY OF LIFE - REALISATION OF TRUTH

After realizing the truth about the Omnipotent Space as the Base & Cause for the vast Universe, as explained above, I wish to continue to write about the 'human beings, their way of life and realization of truth, etc.

CHAPTER FOURTEEN

AT THE OUTSET, I SALUTE TO ALL THE GREAT LEGENDS BORN ON THE PLANET EARTH.

i) As you are aware, our planet Earth is covered by $3/4^{th}$of water and $1/4^{th}$by the land. On this land, now there are 195 countries all over the world with a total population of about eight billion as of now (2019 AD).

ii) There were some people, one in millions, who made strenuous efforts and great research into   all the material things in the universe to explore many unknown, visible and invisible things, by their intelligent study through reasoning, logic, cause and effect. Many such legends sacrificed their entire lives in inventing/ discovering   many things through their   continuous research, dedication and commitment for the welfare of the human society. Presently, we are in the   Electronic Era, which has changed completely the way of life of the present generation of the human society in the world.

iii)My salutations to all those superhuman beings, who sacrificed their lives in inventing/discovering/ exploring many things out of the existing 5 elements in the universe, for development of human society, in improving their knowledge in various fields and to lead the happy and peaceful lives. Those legends are scientists, philanthropists, philosophers, educationists, poets, artists, etc in various fields of Arts, Social Sciences and all the Sciences.

CHAPTER FIFTEEN

THE HUMAN BODY

i. Composition of the Human body:
The human body is formed with invisible basic particles of 11 elements viz. oxygen, carbon, hydrogen, nitrogen, calcium, phosphorus, potassium, sulfur, chlorine, magnesium and sodium, which are essential for life.

ii. Composition and Functions of the Brain:
Of all the living things created in the Universe, the human being is treated as a supreme category, because of the super organ 'Brain', which has Mind, Intelligence, and false ego as follows:

The human being has the mysterious organ 'Brain' located safely in the skull of the head, having millions of thin nerves connecting to all the parts of the body through the main spinal cord and getting signals from sensual organs. It is fatty with billions of neurons, which emanate thoughts of Mind & higher mind (Intelligence). This mind has a conscience of good and bad and controls all the sensual organs. It has millions of nerve fibers to communicate through electrochemical signals with a network of millions of nerves called dendrites and axons. It controls all the external organs of the body, which are sensitive. The fatty acids and essential molecules make the brain to function. The neurons from the brain emanate thoughts created by downloading videos and audios of nature and sounds around since its birth and keep them in memory, which is conceived by the **Mind** to take decisions hurriedly and act accordingly with the help of sense organs. The mind is emotional, has full of material thoughts as it perceives from the world for enjoyments, etc and cannot properly decide what to do or what not to do at a particular situation. Whereas, **Intelligence** is the higher mind. It controls the thoughts of the mind by taking up the right decisions conscientiously. All these thoughts of the Mind and Intelligence are not independent. They all emanate from the Knowledge accumulated in the brain.

Hence, the mind is the actual person and not the external forms of face and body, which are only indexes for a particular type of mind and its nature. However, as all are aware, this vital Brain needs the blood to live. This blood is manufactured and supplied by the whole body consisting of both internal and external organs, like machinery. Hence the body and mind are interdependent. These two are again dependent on the necessary ingredients required to form the blood. Thus a total human being is nothing but the composition of millions of particles of non-living atoms, molecules, cells and their sub-particles of the 5 basic elements and 3 subtle elements of the Brain viz Mind, Intelligence & false ego (feeling of 'Me').

iii. False ego:
The human mind is just like a Sim card of a mobile phone. Before birth it is blank. From birth onwards, it starts downloading all audio, video visuals and feelings perceived through its body organs Eyes, Ears, Mouth, Nose, and Skin. As the human body grows on, various feelings, thoughts, opinions, intellectual ideas, etc are born in it. The legends have invented various languages for communicating their ideas and interacting with each other in human society. Hence every mind firmly feels "I am/Me".

If there is no language, the mind feels nothing except, hearing sounds, tasting various items, smelling various smells, seeing various colors, feeling touching of air and other things to the body skin, and acting with hands, legs, body. All these feelings are created in the system of the brain. Whereas, all these particles of the brain and body do function only due to the presence of the **Inner Space**. Hence, the feeling of 'existence' is felt. But, the brain feels the existence as 'I am/Me', which is called a false ego.

Thus, Mind, Intelligence & false-ego are the 3 invisible subtle energies emanating from the brain. This research process is like a meditation on a particular subject. If you imagine and concentrate totally on the Inner Space, the thoughts of your mind are steadily superseded by your intelligence, duly destroying the false ego, then all these 3 subtle energies automatically merge into the **Infinite Space** eternally, as there is no further base for Space. Consequently, all the elements of the body stop working and die. The **Inner Space** remains united with the **Infinite Space throughout** the Universe intact and undisturbed.

iv. Minds vary from person to person:
In this connection, it is a quintessential point to note that, the character, modus operandi and mental ability varies from person to person, because it depends on the downloaded data of various videos and audios since their childhood at their places of growth, culture, and habits of the people of the society, circumstances prevailing, etc. Thus, depending on these data, the natures of minds vary from person to person. The offspring of the humans continue to have at least some of the characters, expressions, similarities as per their respective Genes & DNA, which exist in the nuclei of cells, carried over from their parents and ancestors.

v. Stages of life - Childhood, Youth & Old age:
After birth, every human being start enjoying childhood, by seeing the colorful nature around - Sun, Sky, Moon, Stars, Rainbows, trees, colorful flowers, birds, animals, people, family members, school, temples, friends, games, fun and frolic, etc, etc.

in their native place. All these things are downloaded into the child's brain. As they grow, their desires will be increasing for various toys, clothes, eatables, etc.

Later, they enjoy youthful life – education, sports, games, fine arts, employment, falling in love, marriage, wife, children, relatives, friends, society, many attachments, across various places of the world, they lived. As they grow, their goals, desires will be increasing endlessly. They go on experiencing the happy moments of life and sometimes sorrowful moments of life. However, they always prefer to enjoy life as long as they are healthy. Though they are aware of the old age and inescapable death on one day or the other, they decline to think about the same and continue to live their full life experiencing joyful and sorrowful moments with compromise. All these experiences will be downloaded in the brain in the form of audios and videos.

Then, Old age comes and they start recollecting the memories of earlier events of life. The body becomes weak, their life is restricted to the home, suffer an illness, lead life worshipping their respective gods, visiting sacred temples, and finally become bedridden until they take their last breath, which is inevitable. All the downloaded audios and videos to vanish along with the brain and body.

vi. What happens after death:

After death, all the gross material particles of the human body merge into their respective 5 elements – Earth, Water, Fire, Air & Ether as non-living particles. The remaining subtle forms of energies viz thoughts of mind, intelligence and false ego, merge into the Infinite Space.

vii. Fear of Death:

By nature, all the living things have been protecting themselves from the other wild animals. Actually, it is the fear of unexpected death on account of attacking of wild animals, natural calamities like floods, storms, cyclones, earthquakes, etc. Fear of death, created in the minds of First humans, made them to pray as their protectors – Sun god, other gods, a king, a lord, a deity, a creator, a supreme power, etc. Later, they started all the elements as personified gods and started worshipping them to protect their lives. Thus the concept of God/Power/Creator arose in the minds of primordial human beings. Later, this had resulted in formation of various religions, principles and practices of worshipping their respective Gods, traditional way of living, doing good deeds, etc, duly following their respective scriptures, preaching of their ancestors and teachers. Thus, after a long period of time, the humans have become civilized, educated, intelligent, making research into various philosophical themes and scientific theories, in search of the truth and goals of their lives.

CHAPTER SIXTEEN

THE REALIZATION OF THE TRUTH

The human beings, being desperate about their physical nature of life and consequential death, started their trials and practices to know the truth and goal of their life. They have been continuing to hear the discourses of their religious teachers, saints, pundits, philosophers, read their books, practice meditation, Yoga, etc., since decades, to know how the universal things are created and what is the existence of life they are feeling in their mind and body, where would they go after death, etc. Hence, to know the ultimate truth, one has to study from the beginning all about the generation of subtlest materials, which formed into a big universe.

No thing can exist without the Omnipotent Space, which is the Base and also the Cause for the generation of everything in the universe. If, one could perceive the Omnipotent Space in every particle of an object, he can also worship his own god with any form, name or color, duly appreciating/respecting the gods/ personified gods of other human beings equally. This practice would enable every human being to observe Oneness in the universe.

Firmly decide and understand that the Inner Space is the life and consciousness for all the living things according to their properties. Similarly the Inner Space is also occupied in the non-living things, but they don't have life and consciousness according to their properties. Therefore, we can understand that the Space exists in both the living and non-living things to help energize and facilitate their sub-particles to perform their respective duties according to their individual properties and mechanisms. For example, Electricity supplied to various appliances is the cause for their functioning according to their mechanisms – Fan, Iron box, Bulb, Grinder, etc.

As stated already, 99.9999999% of human body is occupied by the Inner Space. His whole body and mind altogether is an ignorable part. He must therefore be humble. He must think about his first subtle state when he was born. He is nothing but his Brain. He must delete all the thoughts recorded in his brain during the period of his life or totally ignore them. He must know that the feeling of his presence is nothing but the presence of Infinite Space totally occupied in his body and mind and also all over the universe. The realization of truth is nothing but to know that the Infinite Space is the base and the cause for the creation of all the universal things (both living and non-living) and it also exists in all these things. **One has to realize and experience that the Inner Space within his own body is not separate from the Infinite Space in the Universe.**

CHAPTER SEVENTEEN

How to Approach towards Realization of the Truth?

i. The Space cannot be seen or known by the limited organs of the human body. But at the same time every human mind understands the Space as infinite all over the universe, logically. Space cannot be understood separately from you, because all your body, mind and intelligence are existing only because of the Space in and out. Only deep meditation into the Space leads to higher levels of intelligence and finally the subtleness of the intelligence go on enhancing, till it merges into the infinite space, duly vanishing the false Ego instantaneously.

ii. As more than 99.9999% of the body , brain, mind and intelligence is filled with Inner Space, and also due to the downloaded videos and audios in the brain, the false Ego "I am" is created in the brain. This 'false ego' is only the result of subtlest waves of neurons in the brain. When brain is dead the neurons stop to rise and there will be no thoughts or feelings. Once it is practically realized that the **Inner Space of the body is not separate from the Infinite Space** all over the universe, the memories of the brain will be cleaned, thus destroying the false ego, resulting in immersion of all these subtle material particles in to the Omnipotent Space and the gross material body particles in to their respective gross elements.

CHAPTER EIGHTEEN

HOW TO LIVE AFTER REALIZING THE ABOVE TRUTH

Thus, after realizing the facts about the creation of the Universe by the Infinite Space, as explained above, humans will get full wisdom and energy, which enlighten and get them out of the desperation and distress caused by the temporary life. This energy and wisdom should be utilized for doing good deeds to be happy and to develop the human society. As we are aware, many noble, good and superhuman beings, who sacrificed their lives for the well being of the human society, are treated as demigods /gods by the people of the world. As long as we live in this world, we should be humble, noble and kindhearted towards the destitute and follow the preaching of those super human beings. We must perform all our duties and obligatory functions for the welfare, development and happy living of the human society. We can follow our religious practices as usual. We must also be law abiding, moral, loyal, patriotic citizens of our country. Be happy and let others be happy. This is the purpose of your birth in this material world. However, you can also choose to do the following:

i. Whenever you wish to be relieved from the stress and strain of the mundane life, start meditating upon the Space within you,

38

visualize the same as boundless energy all over the Universe and steadily concentrate. It is out of regular practice, commitment and devotion, one can perceive the Space. Just go on imagining that everything you see is occupied by 99.9999999% of Space and whatever you see is an ignorable portion of material particles, which look solid because of the orbiting of the electrons around the nuclei, etc, as already explained in previous paragraphs. Firstly, you try to observe the Space between any two objects intuitionally, slowly your eyes stop to see those objects even without closing your eyes.

Secondly, close your eyes and try to see the depth of the darkness, the more you concentrate upon any one of the thousands of tiny white dots, you will be able to see wider and wider dark space. You can stop there, and use your intelligence and logic to imagine the infiniteness of the Space all over the universe, inclusive of your body. Continuation of this practice regularly, keeps you happy and neutral in all the good and bad situations that arise in your life and makes you feel Oneness of all the things (living and non-living) in the Universe. This is the triumph over the depression within you which is caused by your unsubstantial, ignorable, temporary nature of life.

ii. Another way of meditation is to concentrate upon the activities of various internal and external organs of your whole body, which are enabled by the Inner Space viz respiration through lungs, heart beating, blood circulation, thoughts emanating from your brain, movement of millions of useful bacteria in your body, hearing through ears, etc. Without the Inner Space, there will be no movement of a tiniest particle of your whole body and mind, which results in total death of the body. You can realize now, that the so called Life or consciousness in all the living things, is nothing but the Omnipotent Space itself, which is right within you and everywhere in the universe.

iii. This realization of Oneness of the Omnipotent Space in all the universal things, inclusive of you, is enough for you to lead your physical life peacefully with more wisdom, vigor and happiness along with your near and dear and the whole human society, enjoying the beautiful, colorful and wonderful nature of the universe, doing all good deeds for the human society, without any despair or distress under all the circumstances of your life.

iv. Finally, whenever you become exhausted and disgusted with all the mundane activities of life, you can, at your own choice, put more efforts to experience the unity of **Inner Space** within you and the **Infinite Space** in the Universe, thus achieving the salvation from the material life.

v. Even though you could not experience the Oneness of the **Omnipotent Space** in the universe, don't worry, your inclined mind at this final stage is enough to get detached from the temporary material world and live happily and blissfully with wisdom in all the situations during the tenure of your life on our beloved planet, Earth.

Gist Of The Book

The infinite Space occupies 99.9999999% of every atom of both living and non-living things in the Universe. All the universe has been created by the Space, from the Space and of the Space energy. Space is the base and the cause for all these universal particles to generate from, to stay in, to grow and finally immerse in the Space when they become weak and die.

All the things in the universe have been generated in the form of invisible and subtle energy particles and later grow up to a level of visible objects, composed with millions of particles of various elements, as is evident from our birth itself.

Omnipotent Space, occupied in the human mind and body, is the cause for the feeling of self existence, life, consciousness and functioning of the whole parts of the body and mind. Without Space the human being becomes a lifeless doll. One's goal should be to realize and experience the Unity of Space within him(Inner Space) & all over the Universe(Infinite Space) - the Oneness of the Omnipotent Space.

Until then, one must lead an humble, noble, happy and peaceful life, doing all good deeds, loving everything in the nature, following their individual religious faiths, practices and principles, duly keeping in mind the Oneness of the Omnipotent Space within him/ her and in all the things in the universe.

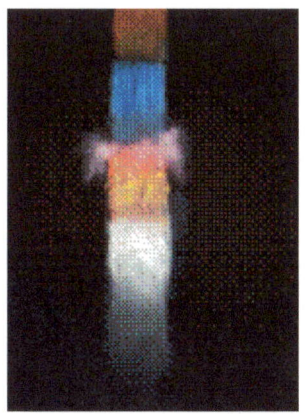

Merger of 5 Elements from Gross to subtle levels -
Earth into Water, Water into Fire, Fire into air, Air
into Ether.

*Merger of 3 subtle elements of the Human Brain -
Mind, Intelligence & False Ego into the infinite
Space, through meditation over the unity of Inner
Space & Infinite Space.*

Formation Of The Universe

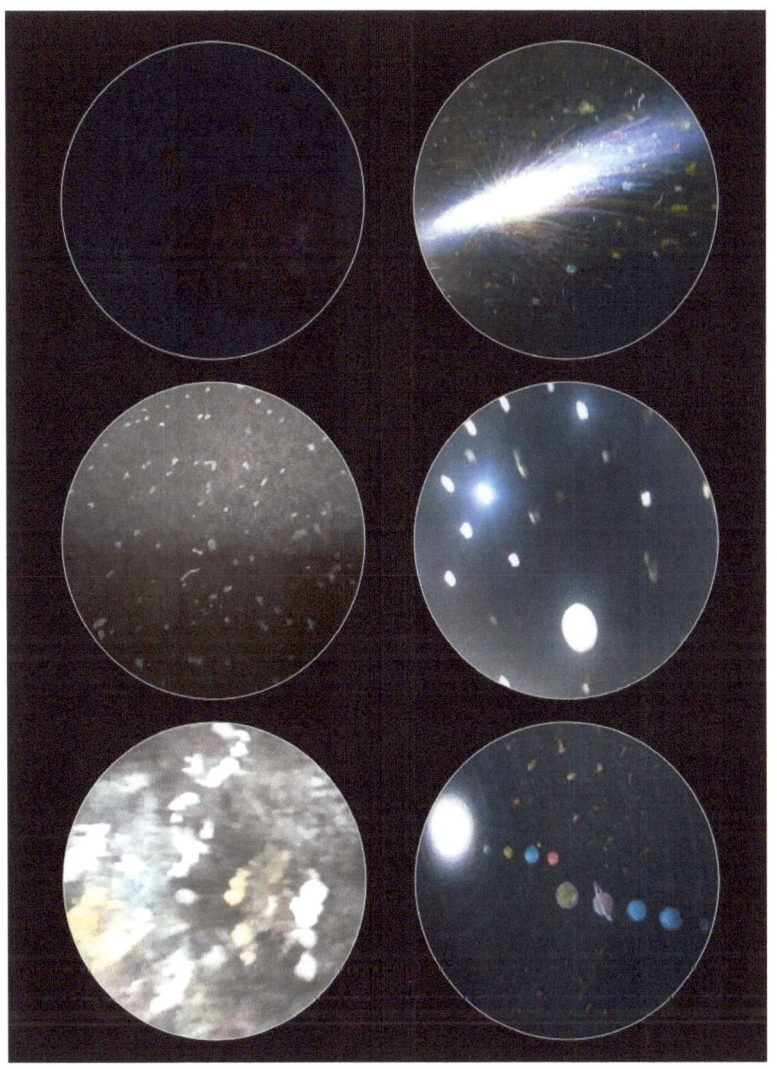